Measu

by Ann Corcorane

Consultant:
Adria F. Klein, Ph.D.
California State University, San Bernardino

capstone
classroom
Heinemann Raintree • Red Brick Learning
division of Capstone

We measure to find out how long.

A tape measure tells us how long.

We measure to find out how heavy.

A scale tells us how heavy.

We measure to find out how far.

Signs tell us how far.

We measure to find out how tall.

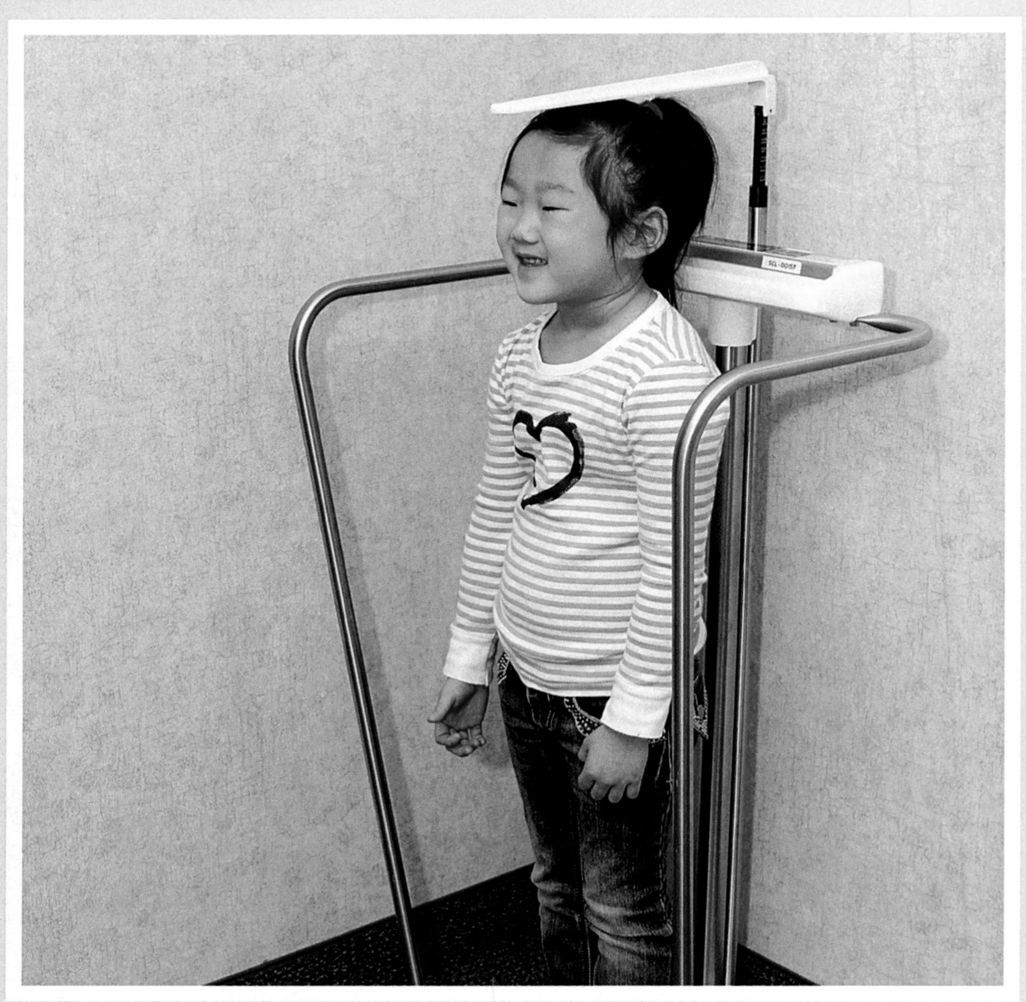

A ruler tells us how tall.

We measure to find out how fast.

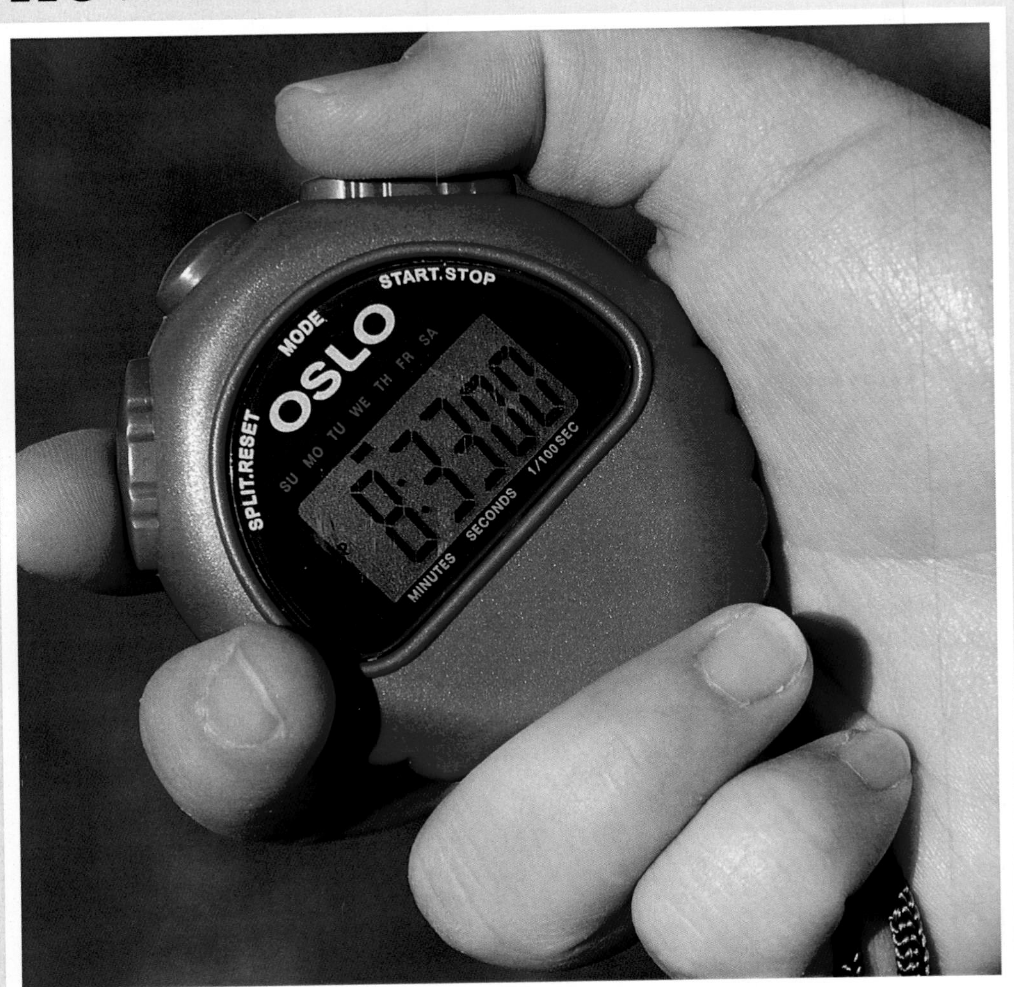

A speedometer tells us how fast.

We measure to find out how much.

A measuring cup tells us how much.

We measure to find out how hot.

A thermometer tells us how hot.

We measure every day.